Abdelhafid Mimouni

Hydrogen sulfide: Landfill risks

Abdelhafid Mimouni

Hydrogen sulfide: Landfill risks

ScienciaScripts

Imprint

Any brand names and product names mentioned in this book are subject to trademark, brand or patent protection and are trademarks or registered trademarks of their respective holders. The use of brand names, product names, common names, trade names, product descriptions etc. even without a particular marking in this work is in no way to be construed to mean that such names may be regarded as unrestricted in respect of trademark and brand protection legislation and could thus be used by anyone.

Cover image: www.ingimage.com

This book is a translation from the original published under ISBN 978-620-6-72109-3.

Publisher:
Sciencia Scripts
is a trademark of
Dodo Books Indian Ocean Ltd. and OmniScriptum S.R.L publishing group

120 High Road, East Finchley, London, N2 9ED, United Kingdom
Str. Armeneasca 28/1, office 1, Chisinau MD-2012, Republic of Moldova, Europe
Printed at: see last page
ISBN: 978-620-8-05737-4

HYDROGEN SULFIDE: LANDFILL RISKS

AUTHOR

Dr. Abdelhafid Mimouni
An independent researcher in bioinorganic chemistry, Dr Mimouni is
an expert in macromolecular synthesis and characterisation. He
obtained his PhD in Chemistry from the University of Paris XII in
1997, after a Diplôme des Études Approfondies in Bioinorganic
Systems from the University of Paris XI in 1993, where he also
obtained his Licence and Maîtrise in Chemistry.

SUMMARY

This book highlights the dangers of toxic chemicals, with a particular focus on hydrogen sulphide (HS2), a noxious gas often emitted from landfill sites. It explores waste management, degradation processes and the release of toxins such as HS2 into the environment. Through an analysis of leachates, this book shows how these fluids contaminate soil and groundwater, seriously affecting public health. The text also discusses the acute and chronic effects of exposure to SH2, the associated pathologies, and the importance of appropriate safety measures for landfill workers. In conclusion, it proposes strengthened policies and innovative techniques for sustainable waste management, aimed at reducing the risks associated with toxic substances and protecting human and environmental health.

4

SUMMARY

INTRODUCTION

Context and objectives of the book

Waste management is one of the most pressing environmental challenges of our time. Although waste is an inevitable consequence of human activity, its inadequate management can lead to serious risks to public health and the environment. This book aims to sound the alarm about the toxicity of chemicals released by waste, particularly those stored in public landfill sites. The aim is to make the public, decision-makers and waste management professionals aware of the insidious dangers that these substances can represent for people living near landfill sites.The book explores not only the toxicity of chemicals, but also their structure, their lethal dose, and the mechanisms by which they contaminate soil and groundwater. By exposing the health risks associated with poor waste management, this book calls for collective awareness and concrete action to protect vulnerable communities.

Overview of waste and landfills

Waste comes from many sources: domestic, industrial, agricultural and commercial. Each of these sources generates distinct types of waste, with varying chemical compositions and specific environmental impacts. Landfill sites, often seen as temporary solutions, have become permanent repositories for a large quantity of waste. of waste. Although regulated, these sites are sometimes sources of pollution due to the release of toxic substances over time. This book begins with an overview of the different types of waste and their sources. It then looks at how landfills operate, the processes by which materials degrade, and how these processes can lead to the release of hazardous chemical compounds. By providing a clear perspective on these issues, this introduction lays the foundations for a more in-depth exploration of the risks and the measures needed to mitigate them in subsequent chapters.

CHAPTER 1
TOXIC CHEMICALS

Chemical Structure and Toxicity

Toxic chemicals are ubiquitous in our environment, originating from a variety of sources such as industrial activities, household products and agricultural waste. Their dangerousness is often linked to their chemical structure, which determines how they interact with biological systems. These compounds include organic and inorganic substances, each with specific toxic mechanisms. For example, benzene, an aromatic hydrocarbon commonly used as an industrial solvent, has a ring structure that favours its absorption by cell membranes. Once in the body, benzene can cause genetic mutations by interfering with DNA, potentially leading to cancer. Another example is lead, a heavy metal whose toxicity results from its ability to substitute for calcium in bones and nerve cells, thereby disrupting a number of biological functions. Analysis of the chemical structure of these substances reveals specific functional groups responsible for their reactivity and toxicity. Halogenated hydrocarbons, for example, have carbon-halogen bonds that make these compounds resistant to biological degradation, increasing their persistence in the environment and their toxic potential.

Lethal Dose (LD50) and Mechanisms of Action

The lethal dose 50 (LD50) is a crucial measure for assessing the toxicity of a chemical. It corresponds to the dose required to kill 50% of the subjects in a test population, generally laboratory animals. This measure is often expressed in milligrams of substance per kilogram of body weight (mg/kg). The mechanisms of action of toxic chemicals vary according to their chemical nature. For example, sodium cyanide (NaCN) is an enzyme inhibitor that works by blocking cytochrome c oxidase, an essential enzyme in the cellular respiration process. By inhibiting this enzyme, cyanide prevents cells from producing ATP, leading to rapid cell death and, in high doses, death of the individual.

Another example is mercury (Hg), which is particularly toxic in its organic form, methylmercury. This compound easily crosses the blood-brain barrier, where it disrupts the functioning of the central nervous system, leading to irreversible neurological damage. The toxicity of mercury is also exacerbated by its accumulation in the food chain, seriously affecting human populations who regularly consume contaminated fish.

Sources and Origins of Chemicals

Toxic chemicals come from a wide range of sources, which can be divided into three broad categories: industrial, domestic and agricultural.

1. **Industry**: Industrial activities are one of the main sources of toxic chemicals. Oil refineries, chemical plants and manufacturing

industries release a wide variety of toxic substances into the environment. These emissions can take the form of solid waste, liquid effluents or gaseous discharges. Volatile organic compounds (VOCs), dioxins, and heavy metals such as mercury and lead are some common examples.

2. **Household products**: In our homes, many common household products contain potentially toxic chemicals. Cleaners, paints, solvents and even some cosmetics can contain dangerous substances such as ammonia, formaldehyde and phthalates. Exposure to these products, even in small quantities, can cause irritation, allergies and other adverse health effects.

3. **Agriculture**: The agricultural sector is also a major source of toxic chemicals, particularly through the use of pesticides, herbicides and fungicides. These products are designed to kill or repel pests, but they can also have toxic effects on humans, especially when they contaminate soil and groundwater. DDT, although now banned in many countries, is a historical example of a highly toxic pesticide that has had devastating effects on the environment and human health.

CHAPTER 2

TOXIC CHEMICALS; THE CASE OF HYDROGEN SULPHIDE

Chemical Structure and Toxicity

Hydrogen sulphide (H2S) is a colourless, flammable gas with a characteristic rotten egg odour in low concentrations. Its simple chemical structure, consisting of two hydrogen atoms linked to a sulphur atom, facilitates its diffusion in the air and poses potential risks in various environmental conditions. H2S is formed naturally by the anaerobic decomposition of organic matter, and is also emitted by industrial sources such as oil refineries, wastewater treatment plants and landfill sites. In these environments, H2S can accumulate to toxic levels, representing a significant hazard to human health.

H2S toxicity results from its interaction with cellular enzyme systems. It inhibits cytochrome c oxidase, a key enzyme in the mitochondrial respiratory chain, a mechanism of action similar to that of cyanide. This inhibition prevents the production of ATP, leading to cellular hypoxia and, in high concentrations, rapid organ failure. Acute exposure to H2S can cause eye and respiratory tract irritation, headaches, nausea and, in extreme cases, death by asphyxiation within minutes.

H2S is particularly dangerous because it can rapidly desensitise olfactory receptors. As a result, exposed people may not detect its odour after an initial brief exposure, increasing the risk of overexposure. accidental inhalation. Concentrations above 100 ppm are immediately dangerous to life and health (IDLH), and concentrations of 300 ppm can be fatal within a few breaths.

Lethal Dose (LD50) and Mechanisms of Action

The lethal dose 50 (LD50) of hydrogen sulphide varies according to species and route of exposure. In rats, the LD50 by inhalation is approximately 444 ppm over four hours, reflecting the high toxicity of H2S at relatively low concentrations. The main mechanism of action of H2S is the inhibition of cytochrome c oxidase in the mitochondria. This enzyme is essential for the transfer of electrons in the respiratory chain, a vital process for the production of ATP. By blocking this enzyme, H2S causes an immediate cessation of cellular respiration, a build-up of lactate, metabolic acidosis and ultimately cell death. This rapid action explains why accidents involving H2S can be sudden and fatal.

H2S readily penetrates cell membranes, rapidly reaching intracellular targets. In addition, its ability to desensitise human olfactory receptors makes it particularly dangerous, as individuals may not perceive its presence after initial exposure.

Sources and origins of hydrogen sulphide

Hydrogen sulphide is produced by both natural and man-made processes:

1. **Natural sources**: H2S is generated by the anaerobic decomposition of organic matter in marshes, wetlands and marine sediments, and is present in volcanic gases and hot springs. Although these sources are important, they are generally localised and less likely to present a risk to human populations.

2. **Anthropogenic sources**: Human activities are a major source of H2S, particularly in industrial sectors. Oil refineries, pulp and paper mills and sewage treatment plants emit high concentrations of H2S. In public landfill sites, H2S is released during the anaerobic degradation of organic waste, reaching dangerous levels in poorly ventilated areas.

3. **Laboratory and industrial use**: H2S is used in certain industrial applications, such as the production of elemental sulphur or sulphur compounds, and in the laboratory for chemical synthesis processes. Its handling requires strict precautions due to its high toxicity.

CHAPTER 3
LANDFILL SITES AND THEIR MANAGEMENT

Landfill operation

Landfill sites are facilities where various types of waste are deposited, ranging from household waste to industrial and agricultural waste. Their management involves several stages designed to minimise the environmental and health impact of the waste deposited.

1. **Waste collection and transport**: Waste is collected from a variety of sources, including households, industries, commercial establishments and farms. It is then transported to landfill sites, which are often located on the outskirts of urban areas.

2. **Waste classification and treatment**: On arrival at the site, waste can be classified into different categories: organic, inorganic, toxic, recyclable, etc. Recyclable materials are generally separated and sent to recycling centres. Recyclable materials are generally separated and sent to recycling centres, while organic and inorganic waste is landfilled.

3. **Landfill methods**: Non-recyclable waste is generally buried in specific cells, covered with layers of soil to limit the dispersion of pollutants. Modern landfills are often equipped with drainage systems to collect leachates, potentially toxic liquids resulting from the infiltration of water through the waste.

4. **Monitoring and managing gases**: Landfill sites produce gases as a result of the anaerobic degradation of organic waste. These gases

include, These include methane (CH4), carbon dioxide (CO2) and hydrogen sulphide (H2S). Landfill sites must be equipped with gas collection and treatment systems to prevent gases from being released into the atmosphere.

The Process of Degradation and Release of Chemicals

Waste deposited in landfill sites undergoes degradation processes that can lead to the release of potentially toxic chemicals, including hydrogen sulphide (H2S).

1. **Anaerobic degradation of organic waste**: Organic waste, such as food scraps, agricultural residues and other biological matter, is degraded by micro-organisms in the absence of oxygen. This fermentation process produces gases such as methane and hydrogen sulphide, the latter being of particular concern because of its high toxicity.

2. **Leachate formation**: When water (rain, seepage) passes through layers of waste, it becomes loaded with dissolved substances and forms what is known as leachate. This liquid can contain a variety of toxic chemicals, including heavy metals, volatile organic compounds and dissolved gases such as H2S. If leachate drainage and treatment systems are not properly installed or maintained, these substances can contaminate soil and groundwater, posing a major risk to public health.

3. **Release and Dispersion of Toxic Gases**: H2S, as a product of anaerobic degradation, can be released from landfills, especially in

areas where gas capture systems are ineffective or non-existent. Because of its high toxicity, even at low concentrations, the presence of H2S in ambient air constitutes a serious hazard for people living near landfill sites. The effects of chronic exposure include respiratory problems, headaches, eye irritation and, in extreme cases, loss of consciousness or death.

4. **Impact of Environmental Conditions**: Environmental conditions, such as temperature, humidity, and the presence of organic matter, influence the rate of waste degradation and the amount of gas and leachate produced. For example, a landfill located in a hot, humid region could produce more H2S than a landfill in a cold, dry region.

CHAPTER 4
LEACHATES: COMPOSITION AND IMPACT

Chemical composition of leachates

Leachates are highly polluting liquids that form when water passes through waste accumulated in landfill sites. This process results in the dissolution and collection of various chemical compounds, some of which are extremely toxic to the environment and human health. The chemical composition of leachates is complex and variable, depending on the nature of the waste, weather conditions and landfill management.

1. Heavy metals

Leachates often contain heavy metals such as lead, mercury, cadmium and arsenic. These metals mainly come from electronic waste, batteries and other industrial products. Heavy metals are of particular concern because of their persistence in the environment and their ability to accumulate in living organisms, causing serious toxic effects such as cancer, neurological disorders and kidney dysfunction.

2. Volatile Organic Compounds (VOCs)

Volatile organic compounds such as benzene, toluene and xylene are commonly present in leachates. These substances, derived from petroleum products, industrial solvents and household waste, are known for their carcinogenic properties and harmful effects on the respiratory system.

3. Hydrogen sulphide (H2S)

Hydrogen sulphide (H2S) is a toxic gas that can also be found dissolved in leachates. This compound comes mainly from the anaerobic degradation of organic matter, such as food and agricultural waste, present in landfill sites. H2S is extremely dangerous, even in low concentrations, and can cause severe respiratory damage, eye irritation, and in extreme cases, prolonged exposure can be fatal.

4. Nutrients

Leachates often contain high concentrations of nutrients such as nitrates and phosphates, derived from food waste and fertilisers. Although these elements are essential to life, their excess can lead to eutrophication of water bodies, resulting in serious ecological imbalances.

5. Pathogens

In addition to chemical contaminants, leachates can contain pathogens such as bacteria, viruses and parasites from organic and medical waste. These micro-organisms can contaminate soil and water, posing a risk to human and animal health.

Soil and groundwater contamination mechanisms

Leachates are a major source of soil and groundwater pollution. The mechanisms of contamination are mainly linked to the infiltration of these toxic liquids through the layers of soil beneath landfill sites.

1. Infiltration and Dispersion

When leachate seeps into the ground, it can carry with it a multitude of contaminants, which are then dispersed into the lower layers of the soil. This process can lead to widespread pollution, affecting not only the immediate areas around landfills, but also more distant regions, depending on local geology and groundwater movements.

2. Underground Water Migration

Groundwater is particularly vulnerable to contamination by leachates. Aquifers, which supply drinking water to many populations, can be seriously affected. Heavy metals, VOCs, H2S and other toxins present in leachates can penetrate aquifers, making the water unfit for consumption and posing risks to public health.

3. Accumulation in the Food Chain

Contaminants in leachates can accumulate in soils and then be absorbed by plants, entering the food chain. Heavy metals, in particular, can be absorbed by crops, exposing human and animal populations to the risks associated with consuming contaminated products.

Case Studies

1. Love Canal, New York, United States

One of the most famous cases of leachate contamination is Love Canal, a residential area built on a former chemical landfill site. In the 1970s, thousands of tonnes of waste were dumped in this area. toxic substances, including heavy metals and volatile organic compounds, began to seep into the soil and groundwater, leading to a public health crisis. Residents suffered from a range of health problems, including cancer, birth defects and chronic illnesses. The incident led to the evacuation of residents and the creation of the Superfund programme in the United States to clean up contaminated sites.

2. Lekwena Landfill, South Africa

In South Africa, the Lekwena landfill also posed serious environmental problems. Leachates rich in H2S and other toxic compounds have contaminated groundwater, affecting surrounding villages. Residents reported frequent cases of respiratory illness, headaches and other symptoms linked to exposure to H2S. The contamination has also affected local agriculture, compromising crops and water sources.

3. ToxiCo depot, Italy

In Italy, the ToxiCo dump, an illegal landfill site, released massive quantities of leachate containing heavy metals and H2S into the environment. The leachate penetrated local aquifers, leading to widespread contamination of drinking water supplies. Local residents have suffered serious health problems, including cardiovascular disease and respiratory ailments. This case highlighted the importance of strict regulation and monitoring of landfill sites.

CHAPTER 5
EFFECTS ON PUBLIC HEALTH

Acute and Chronic Toxicity

Exposure to toxic chemicals, such as those present in landfill leachate, can cause a range of health effects, from acute toxicity to serious chronic illness. Acute toxicity usually occurs immediately or shortly after exposure to a high concentration of a toxic substance. Common symptoms include nausea, vomiting, headaches, skin and respiratory irritation, and in severe cases, loss of consciousness or death. For example, inhalation of hydrogen sulphide (HS2) can cause breathing difficulties, rapid loss of consciousness and, in high concentrations, death.Chronic toxicity, on the other hand, results from prolonged exposure to low doses of toxic substances, often over many years. Effects can include cancer, cardiovascular disease, neurological disorders and kidney dysfunction. SH2, even at low levels, can cause chronic respiratory problems, persistent eye irritation and effects on the central nervous system.

Exposure-related diseases

Exposure to toxic chemicals in leachates, particularly in areas close to landfill sites, is associated with a wide range of pathologies. Cancerous diseases are of particular concern, as many Compounds present in leachates, such as heavy metals and volatile organic compounds, are proven carcinogens. Benzene, for example, is associated with leukaemia, while cadmium is linked to kidney and lung cancer. Respiratory problems are also common among people living near landfill sites, largely due to the inhalation of toxic gases such as SH2. This gas, even at relatively low concentrations, can cause respiratory tract irritation, asthma and chronic bronchitis. Other long-term effects include cardiovascular disease, often exacerbated by the environmental stress associated with chemical pollution.

Vulnerable populations

Certain populations are particularly vulnerable to the harmful effects of toxic substances present in landfill leachates. Children, because of their small size and ongoing development, absorb proportionately more toxins than adults and are more likely to suffer from developmental disorders, asthma and other chronic illnesses. The elderly, whose immune systems are often weakened, are also at greater risk of developing serious illnesses following prolonged exposure to pollutants.Communities living near landfill sites are exposed to higher levels of contaminants, and residents of these areas often come from This limits their access to adequate healthcare and increases their vulnerability. Landfill workers are also on the front line, as they are exposed to high concentrations of toxic substances on a daily basis, making them particularly prone to occupational illness.

Behaviour of Gases in Landfills

Hydrogen sulphide (HS2) is a toxic gas with a density of around 1.19 kg/m^3 at 0°C and 1 atm. In comparison, carbon dioxide (CO2) is denser, with a density of around 1.98 kg/m³, while oxygen (O2) is slightly lighter than SH2, with a density of 1.43 kg/m³. This difference in density influences the distribution of gases in landfill sites. HS2, being heavier than oxygen but lighter than carbon dioxide, tends to accumulate close to the ground in confined or low-lying environments, increasing the risk of exposure for people working in these conditions.

Toxic gas detection: Focus on SH2

Hydrogen sulphide (HS2) is a colourless, flammable and extremely toxic gas that can be formed in landfill sites from the anaerobic decomposition of organic matter. To prevent the harmful effects of this gas, it is essential to detect it quickly and effectively. Gas detectors play a crucial role in monitoring SH2 levels in the air, particularly in high-risk environments such as landfill sites. There are several types of SH2 detectors, ranging from portable devices to fixed systems installed on landfill sites. Miniature detectors are particularly useful for continuous personal monitoring, enabling workers to react quickly in the event of a gas leak. These devices are often equipped with audible and visual alarms that are triggered when SH2 levels exceed established safety thresholds. Fixed gas detectors, meanwhile, are often connected to ventilation or alarm systems that can automatically activate emergency measures.

Safety measures for workers

Protecting workers at landfill sites is a priority, given the high risk of exposure to toxic chemicals and gases such as SH2. Personal protective equipment (PPE) is essential to minimise these risks. Commonly used PPE includes:

- **Respiratory masks**: Filtering or cartridge masks are used to protect the respiratory tract against the inhalation of toxic gases and airborne particles.
- **Protective Goggles**: These goggles prevent irritation and eye damage caused by gases or chemical splashes.
- **Chemical-resistant gloves**: Gloves made from resistant materials such as neoprene or nitrile protect workers' hands against contact with corrosive or toxic substances.
- **Protective suits**: Tight-fitting suits prevent direct contact with chemicals and leachates, reducing the risk of skin absorption of toxins.

In addition to PPE, regular training of workers in safety protocols and first aid is essential to ensure their protection.

HS2 Mitigation Techniques

Reducing the concentration of SH2 in the landfill environment is essential to protect the health of workers and surrounding populations. Several mitigation techniques can be used:

- **Ventilation**: Installing efficient ventilation systems dilutes and disperses SH2, reducing its concentration in ambient air.

- **Chemical Neutralisation**: The use of chemical compounds such as sodium peroxide can neutralise SH2, transforming this toxic gas into harmless compounds.

- **Filtration Systems**: Activated carbon filters can be used to absorb SH2, preventing its release into the air.

- **Covering waste** : Regular application of layers of soil or other inert materials over waste at landfill sites can reduce the formation of SH2 by limiting the exposure of organic material to the air.

CHAPTER 6
POLICIES AND REGULATIONS

Existing regulatory framework

Waste management and the protection of public health are governed by a range of national and international laws and regulations designed to minimise the risks associated with exposure to toxic substances, including hazardous gases such as hydrogen sulphide (HS2). These laws define the standards for waste disposal, landfill construction and management, as well as the safety measures to be put in place to protect workers and neighbouring populations.For example, regulations in many countries require regular monitoring of leachate and gas emissions from landfill sites. They also require the installation of detection systems for toxic gases such as SH2, and the implementation of safety measures such as ventilation and the use of personal protective equipment (PPE) for workers. At international level, agreements such as the Basel Convention govern the transboundary movement of hazardous waste and its disposal, seeking to minimise the risks to the environment and human health. In addition, bodies such as the World Health Organisation (WHO) provide guidelines for waste management and the prevention of health risks associated with exposure to toxic substances.

Flaws and Limits of Current Regulations

Despite the existence of these regulatory frameworks, a number of gaps and limitations remain, compromising the protection of people and the environment. One of the main weaknesses is the uneven application of laws, often due to limited resources or a lack of political will. In many cases, landfill sites, particularly in developing countries, do not meet minimum safety standards, leading to increased exposure of populations to toxic substances.Another major limitation is the inadequacy of current regulations to deal with new environmental threats, such as the increase in electronic waste or microplastics. These materials can release toxic substances that are not covered by existing regulations. In addition, standards for SH2 emissions from landfill sites are often based on obsolete data, failing to take account of new scientific discoveries about the toxicity of this gas at low concentrations. There is also a lack of continuous monitoring and follow-up of the long-term effects of exposure to leachates and toxic gases, which prevents the rapid identification of risks to public health and the environment.

Suggestions for Improvement

A number of improvements could be made to the existing regulatory frameworks to enhance public protection against the risks associated with landfills and toxic substances:

1. **Strengthening Enforcement**: It is crucial to improve enforcement by ensuring that all landfills, whether public or private, comply with safety standards. This could be achieved through increased inspections and tougher penalties for non-compliance.

2. **Updating standards and regulations**: Standards for emissions of toxic gases such as HS2 need to be revised regularly to reflect the latest scientific knowledge. Regulations should also include specific measures for new types of waste, such as electronic waste and microplastics, which pose new environmental risks.

3. **Promoting Research and Monitoring**: Investing in research into the long-term effects of exposure to toxic substances and improving monitoring systems for leachates and landfill gas emissions. These measures would enable hazards to be identified more quickly and more effective prevention strategies to be put in place.

4. **Improved Detection and Protection Systems**: The use of advanced technologies for the detection of toxic gases, including real-time monitoring systems, should be encouraged. In addition, it is essential to improve access to personal protective equipment for workers, particularly in regions where resources are limited.

5. **Awareness-raising and training**: Raising awareness of health risks among people living near landfill sites and training workers in good safety practices should be stepped up. Public information campaigns and specific training programmes could play a key role in reducing the risks of exposure to toxic substances.

CHAPTER 7

SUSTAINABLE WASTE MANAGEMENT APPROACHES

Advanced Waste Treatment Techniques

Sustainable waste management means adopting advanced techniques to treat and reduce the toxicity of waste. Several technologies are emerging and being perfected to meet this challenge.

1. **Mechanical-Biological Treatment**: This approach combines mechanical processes, such as waste sorting and shredding, with biological methods, such as composting and anaerobic digestion. Mechanical-biological treatment reduces the volume of waste, stabilises organic matter and produces less harmful compost or biogas.

2. **Decontamination technologies**: Decontamination processes, such as thermolysis or advanced oxidation, aim to break down the toxic substances present in waste. Thermolysis uses heat to break down hazardous compounds into less harmful products, while advanced oxidation uses oxidising agents to degrade organic pollutants.

3. **Filtration and Neutralisation Systems**: Activated carbon filters and chemical treatment systems are used to capture and neutralise the toxic substances released by waste. These technologies are particularly useful for treating leachates and gases released by landfill sites.

4. **Waste recovery**: Waste recovery techniques, such as pyrolysis and gasification, transform waste into energy products or reusable materials. Pyrolysis breaks down waste into gas, oil and coal, while

gasification converts waste into synthetic gas that can be used as a source of energy.

Innovative Risk Reduction Solutions

To minimise the risks to public health and the environment, several innovative approaches can be implemented:

1. **Intelligent Waste Management**: The use of intelligent management technologies, such as sensors to monitor pollution levels and data-driven waste management systems, enables more efficient landfill management. The data collected can help optimise waste management operations and identify potential risks in real time.

2. **Circular Economy**: The circular economy approach aims to reduce waste by reusing, repairing and recycling materials. This reduces the amount of waste sent to landfill and minimises the environmental impact of end-of-life products. Business models based on circularity, such as the repair of electronic devices and the recycling of plastic materials, play a key role in this approach.

3. **Innovations in Materials**: The development of alternative materials, such as biodegradable plastics and eco-friendly packaging, is helping to reduce the production of toxic waste. These materials are designed to break down more quickly and have a reduced environmental impact compared to traditional materials.

4. **Waste pre-treatment systems**: The application of pre-treatments, such as automated sorting and separation of contaminants, reduces the toxicity of waste before it is deposited in landfill. This includes sorting

hazardous waste and separating recyclable materials, thereby limiting the amount of harmful substances in the leachate.

The role of public policy and civil society

Sustainable waste management requires close cooperation between governments, non-governmental organisations (NGOs) and civil society. Here's how these stakeholders can contribute:

1. **Public policy**: Governments play a crucial role in establishing and implementing strict regulations on waste management. They can introduce policies to reduce waste at source, encourage recycling and reuse, and support waste management programmes. Financial incentives, such as subsidies for waste treatment technologies or taxes on untreated waste can also encourage more sustainable management practices.

2. **Civil society and NGOs**: Non-governmental organisations and community groups can raise public awareness of waste management issues and promote more responsible behaviour. They can also play a role in implementing clean-up, community composting and small-scale waste management programmes. They can complement government efforts by creating local initiatives and monitoring waste management practices.

3. **Public-Private Partnerships**: Public-private partnerships can foster the development and application of innovative waste management technologies. These collaborations can include research and development, the implementation of new technologies, and the financing of waste management projects.

4. **Community involvement**: Involving local communities in waste management, through selective sorting programmes, educational workshops and waste reduction initiatives, strengthens the effectiveness of waste management policies and improves residents' quality of life.

CONCLUSION

Risk Summary and Call to Action

Over the course of this book, we have explored in detail the dangers associated with toxic chemicals, with a particular focus on hydrogen sulphide (HS2), a particularly noxious gas from landfill sites. We have seen how HS2, emitted during the decomposition of organic matter in landfill sites, can have serious effects on human health and the environment. Its acute toxicity, in particular its corrosive and respiratory effects, as well as its chronic effects, such as irritation of the respiratory tract and increased risks for people living near landfill sites, underline the urgency of the situation.Leachates, produced by waste in landfill sites, are also a major source of pollution, releasing toxic substances, including SH2, into the soil and groundwater. This contamination poses major risks to public health and the environment, exacerbated by the shortcomings and limitations of current regulations.Faced with these challenges, it is imperative that we take concrete steps to reduce the risks associated with waste and HS2. This includes strengthening regulations, adopting advanced treatment technologies, and putting in place effective safety measures for landfill workers. Raising awareness and involving local communities, as well as cooperation between the public and private sectors, are also crucial to addressing these issues.

The future of waste management

The future of waste management is taking shape through innovation and the continuous improvement of management practices. New technologies, such as advanced detection systems for HS2 and innovative treatment methods, will play a key role in reducing the toxicity of waste and minimising the risks to public health.The transition to a circular economy, aimed at reducing waste by reusing and recycling materials, is also a promising direction. This approach not only reduces the amount of waste sent to landfill but also reduces the release of toxic substances such as HS2.

Public policies will have to evolve to incorporate these new realities, with stricter regulations and reinforced control measures. Greater involvement of civil society and businesses in waste management, and the promotion of innovative solutions, will be essential to building a more sustainable and secure future.

In conclusion, waste management, and in particular the management of HS2-related risks, requires concerted and continuous action. Through combined efforts and collective awareness, we can reduce the negative impacts of This will help to reduce the impact of waste on our environment and our health, while moving towards more sustainable and efficient management practices.

GLOSSARY

- **VOCs (Volatile Organic Compounds)** : Organic substances that vaporise easily in the air and can contribute to air pollution.

- **LD50 (Lethal Dose 50)** : The dose of a substance that causes death by 50 % of subjects in a test population.

- **Cytochrome c oxidase**: Key enzyme in the mitochondrial respiratory chain, essential for ATP production in cells. Inhibited by H2S, it prevents energy production in cells.

- **Methylmercury**: Organic form of mercury, highly toxic and capable of accumulating in living organisms, particularly in the nervous system.

- **DDT (Dichlorodiphenyltrichloroethane)**: A once widely used organochlorine insecticide, now banned in many countries due to its high toxicity and persistence in the environment.

- **Hypoxia**: Condition in which the body or a region of the body is deprived of an adequate supply of oxygen.

- **ATP (Adenosine Triphosphate)** : Main energy molecule used by cells to carry out various biological functions.

- **Lactic acidosis**: Accumulation of lactic acid in the body, often due to defective cellular respiration.

- **Homeostasis**: An organism's ability to maintain a stable internal state despite external changes.

- **IDLH (Immediately Dangerous to Life or Health)**: Concentration of a substance in air that poses an immediate danger to life or health if exposed.

- **Anaerobic**: Biological process taking place in the absence of oxygen.

- H2S **(Hydrogen Sulphide)**: Colourless toxic gas with a rotten egg smell, produced by the decomposition of organic matter and certain industrial activities.
- **Leachate**: Liquid resulting from the infiltration of water through waste, containing potentially toxic dissolved substances.
- **Methane** (CH4): Colourless, odourless and flammable gas produced by the decomposition of organic matter in an anaerobic environment.
- **Eutrophication**: Process by which a body of water becomes rich in nutrients, leading to excessive algal growth and a reduction in available oxygen.
- **Aquifer**: Geological formation containing underground water that can be exploited for water supply.
- **Love Canal**: Famous contamination site in New York, symbol of environmental pollution by toxic waste.
- **Superfund**: Programme to clean up sites contaminated by hazardous waste in the United States.

- **Hydrogen Sulphide** (HS2): A colourless, toxic gas, often produced by the decomposition of organic matter in anaerobic environments.
- **Personal Protective Equipment (PPE)**: A set of equipment designed to protect workers against occupational hazards.
- **Regulatory framework**: set of laws and regulations governing a specific area, in this case waste management and public health protection.
- **Basel Convention**: International agreement governing the transboundary movement of hazardous waste and its disposal.

- **Electronic waste**: Waste from end-of-life electronic products, often containing toxic substances such as lead and mercury.
- **Circular Economy**: Economic model aimed at maximising the reuse and recycling of materials throughout the product life cycle, thereby reducing waste.
- **Activated Carbon Filtration**: Purification technology used to capture toxic substances and impurities from gases or liquids.
- **Gasification**: Process of converting waste into synthesis gas by thermal reaction with a gaseous agent.
- **Pyrolysis**: Thermal decomposition of waste in the absence of oxygen, producing gases, oil and coal.

REFERENCES

• Agency for Toxic Substances and Disease Registry (ATSDR). (2019). Toxicological profile for lead. U.S. Department of Health and Human Services.

• Klaassen, C. D. (Ed.). (2013). Casarett and Doull's Toxicology: The Basic Science of Poisons (8th ed.). McGraw-Hill Education.

• Lide, D. R. (2009). CRC Handbook of Chemistry and Physics (90th ed.). CRC Press.

• Snyder, R. (Ed.). (2014). Ethel Browning's Toxicity and Metabolism of Industrial Solvents (2nd ed.). Elsevier.

• WHO (World Health Organization). (2020). Health effects of heavy metals. Retrieved from https://www.who.int/news-room/fact-sheets/detail/heavy- metals

• Beauchamp, R. O., Bus, J. S., Popp, J. A., Boreiko, C. J., & Andjelkovich,D. A. (1984). A critical review of the literature on hydrogen sulfide toxicity. Critical Reviews in Toxicology, 13(1), 25-97.

• National Institute for Occupational Safety and Health (NIOSH). (2022). Hydrogen sulfide: Chemical safety information. U.S. Department of Health and Human Services. Retrieved from https://www.cdc.gov/niosh/npg/npgd0337.html

• Reiffenstein, R. J., Hulbert, W. C., & Roth, S. H. (1992). Toxicology of hydrogen sulfide. Annual Review of Pharmacology and Toxicology, 32(1), 109-134.

• Snyder, R. (Ed.). (2014). Ethel Browning's Toxicity and Metabolism of Industrial Solvents (2nd ed.). Elsevier.

• World Health Organization (WHO). (2003). Hydrogen sulphide: Human health aspects. WHO Press.

• Guidotti, T. L. (1996). Hydrogen sulfide. Occupational Medicine, 46(5), 367-371.

• Guengerich, F. P. (2007). Mechanisms of metal-induced carcinogenesis. Biological Trace Element Research, 119(1), 50-64.

• Williams, A. L., & Jones, J. L. (1998). Environmental contamination from industrial sources: The role of government regulation. Environmental Management, 22(3), 451-460.

• Alloway, B. J., & Ayres, D. C. (1997). Chemical Principles of Environmental Pollution. Chapman & Hall.

• Christensen, T. H., Kjeldsen, P., Bjerg, P. L., Jensen, D. L., Christensen, J. B., Baun, A., ... & Heron, G. (2001). Biogeochemistry of landfill leachate plumes. Applied Geochemistry, 16(7-8), 659-718.

• Love Canal Homeowners Association. (1981). The Love Canal tragedy. Environmental Protection Agency.

• Wicks, C. M., & Troch, P. A. (2001). Hydrogeology of contaminant plumes in granular aquifers. Hydrogeology Journal, 9(5), 508-520.

• ATSDR. (2006). Toxicological Profile for Hydrogen Sulfide. Agency for Toxic Substances and Disease Registry.

• Jacobson, M. Z. (2005). Fundamentals of Atmospheric Modeling. Cambridge University Press.

• NIOSH. (2005). Hydrogen Sulfide (NIOSH Publication No. 2005-149). National Institute for Occupational Safety and Health.

• Zey, J. N. (2001). Hydrogen Sulfide: Occupational Hazards and

Safety. Journal of Occupational Health, 43(5), 309-317.

• French Environment and Energy Management Agency (ADEME). (2017). Waste Management: Regulatory Framework and Applications. Paris: ADEME.

• Basel Convention (1989). Basel Convention on the Control of Transboundary Movements of Hazardous Wastes and their Disposal. United Nations.

• World Health Organization (WHO). (2004). Guidelines for Drinking-water Quality. Geneva: WHO Press.

• Wilson, D. C., Velis, C., & Cheeseman, C. (2006). Role of informal sector recycling in waste management in developing countries. Habitat International, 30(4), 797-808.

• Bove, A., & Balestrieri, L. (2017). Advanced Waste Treatment Technologies. Springer.

• Ellen MacArthur Foundation. (2013). Towards the Circular Economy: Economic and Business Rationale for an Accelerated Transition. Ellen MacArthur Foundation.

• Kinnunen, P., & Suh, S. (2018). Techniques and Technologies for Waste Management and Recycling. CRC Press.

• Zhang, X., & Xu, W. (2020). Innovative Materials for Waste Management. Elsevier.

• Agency for Toxic Substances and Disease Registry (ATSDR). (2019). Toxicological Profile for Hydrogen Sulfide. U.S. Department of Health and Human Services.

I want morebooks!

Buy your books fast and straightforward online - at one of world's fastest growing online book stores! Environmentally sound due to Print-on-Demand technologies.

Buy your books online at
www.morebooks.shop

Kaufen Sie Ihre Bücher schnell und unkompliziert online – auf einer der am schnellsten wachsenden Buchhandelsplattformen weltweit! Dank Print-On-Demand umwelt- und ressourcenschonend produziert.

Bücher schneller online kaufen
www.morebooks.shop

info@omniscriptum.com
www.omniscriptum.com